and Meteors

By Gregory Vogt

www.steck-vaughn.com

OUR UNIVERSE

ISBN 0-7398-3351-0

Printed and bound in the United States of America
1 2 3 4 5 6 7 8 9 W 04 03 02 01 00

Photo Acknowledgments
NASA, cover, title page, 10, 14, 17, 20, 23, 24, 28, 35, 40, 42, 44; USGS/NASA, 39; Hubble Heritage Team (NASA/AURA/STScI), 9, 16; Corbis, 18, 26, 30, 32, 36

Contents

Diagram of a Comet

Type I Tail – made mostly of gas.

Type II Tail – made mostly of dust.

Coma – made of gas and dust.

Nucleus – made up of frozen gas, rock, and dust.

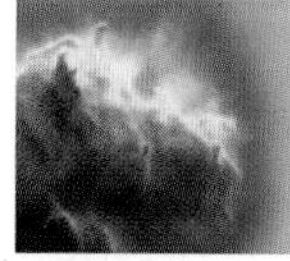

A Quick Look at Asteroids, Comets, and Meteors

What is an asteroid?

An asteroid is a large rock in space that orbits the Sun.

What are asteroids made of?

Asteroids can be made of rock, ice, metal, and minerals.

What is a comet?

A comet is a large ball of ice, dust, and rock that has a tail of glowing gases when it is near the Sun.

What is a comet tail made of?

There are two types of comet tails. One kind of tail is made mostly of dust. The second kind of tail is made mostly of gas.

How many tails can comets have?

Comets can have many tails. One observed comet had six tails.

What is a meteor?

A meteor is a meteoroid that has entered the layer of gas that surrounds Earth. Meteoroids are dust and small rocks in space. Meteors burn up in the layer of gases. This causes a bright flash.

What is a shooting star?

Shooting star is the nickname for a meteor that is streaking across the sky.

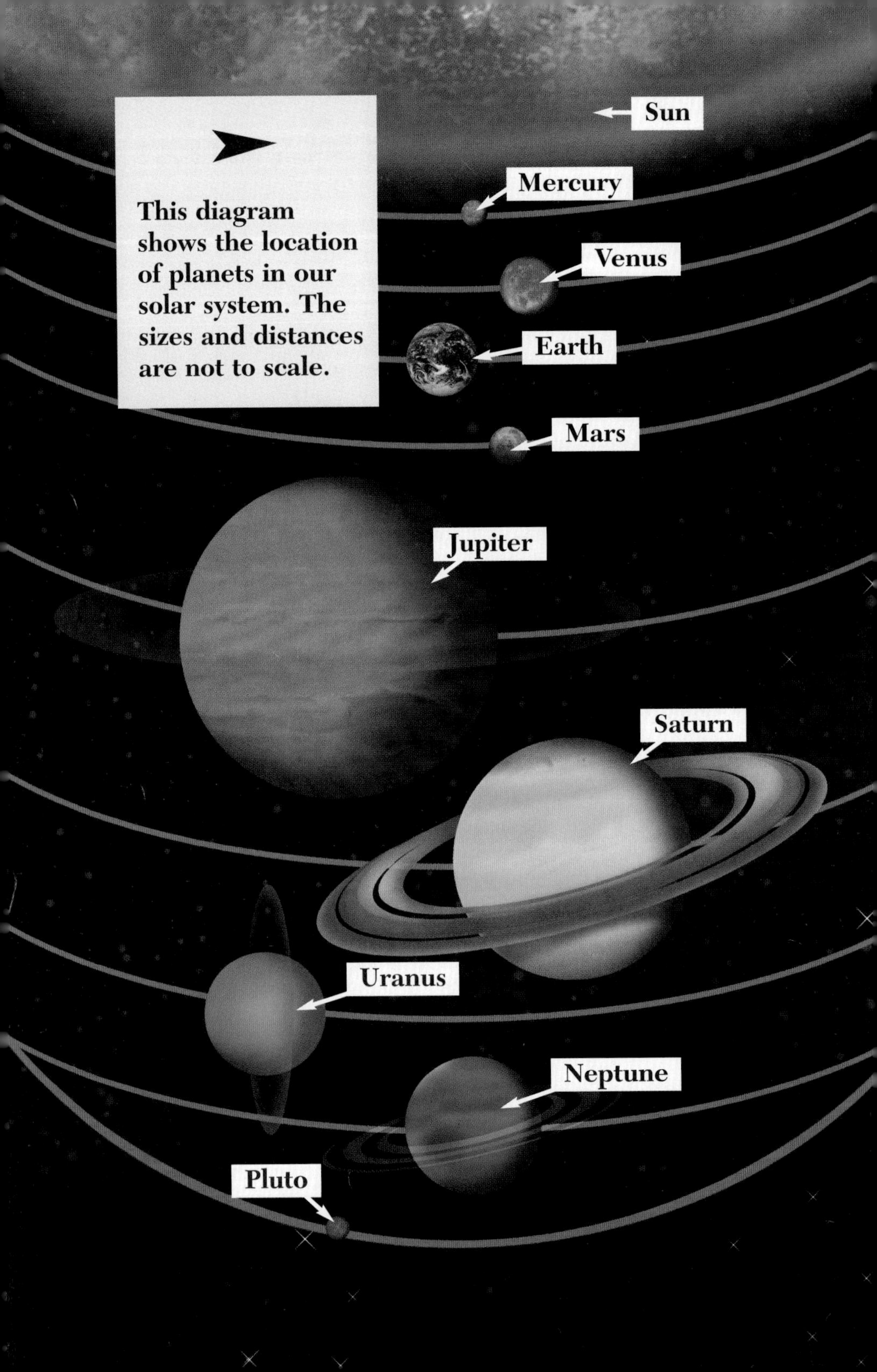
This diagram shows the location of planets in our solar system. The sizes and distances are not to scale.
Sun
Mercury
Venus
Earth
Mars
Jupiter
Saturn
Uranus
Neptune
Pluto

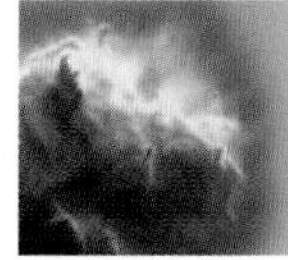

Small Objects in the Solar System

The solar system is made up of the Sun and everything that orbits it. An orbit is a path an object takes as it travels around another object in space. Earth is one of nine planets in our solar system. Most planets have one or more moons orbiting them. The Sun, planets, and moons are the largest objects in the solar system.

Trillions of smaller objects are part of the solar system, too. These objects are large rocks called asteroids, dust and smaller rocks called meteoroids, and comets. A comet is a large ball of ice, dust, and rock that has a tail of glowing gases when it is near the Sun.

Sometimes these smaller objects crash into planets, moons, or the Sun. The crashes break the asteroids, comets, and meteoroids into smaller pieces or completely destroy them.

Solar System Leftovers

Scientists believe the solar system formed billions of years ago in a nebula. A nebula is a huge cloud of gas and dust. One idea is that the gravitational pull of the dust and gas made the nebula slowly collapse. Gravity is a natural force that attracts objects to each other. The more dense, or heavy, an object is, the more it pulls on objects around it. This pull is called an object's gravitational force. The nebula became a flat disk of gas and dust. The disk began to spin.

Most of the gas and dust fell to the center of the nebula and became the Sun. The gas got very hot. The hydrogen gas began to turn into helium. Energy from this process was released as light and heat.

The Sun's gravity pulled the rest of the gas and dust into an orbit around the Sun. Clumps of dust formed as the disk swirled around the Sun. Smaller clumps joined together when they crashed into each other. They became larger clumps. Over millions of years, the clumps became planets and moons.

Some material was left over after the large objects in the solar system formed. This material joined together to form asteroids, comets, and meteoroids. Some meteoroids are small pieces of rock or dust that have broken off from asteroids or comets.

The clumps inside this nebula are new stars and planets forming.

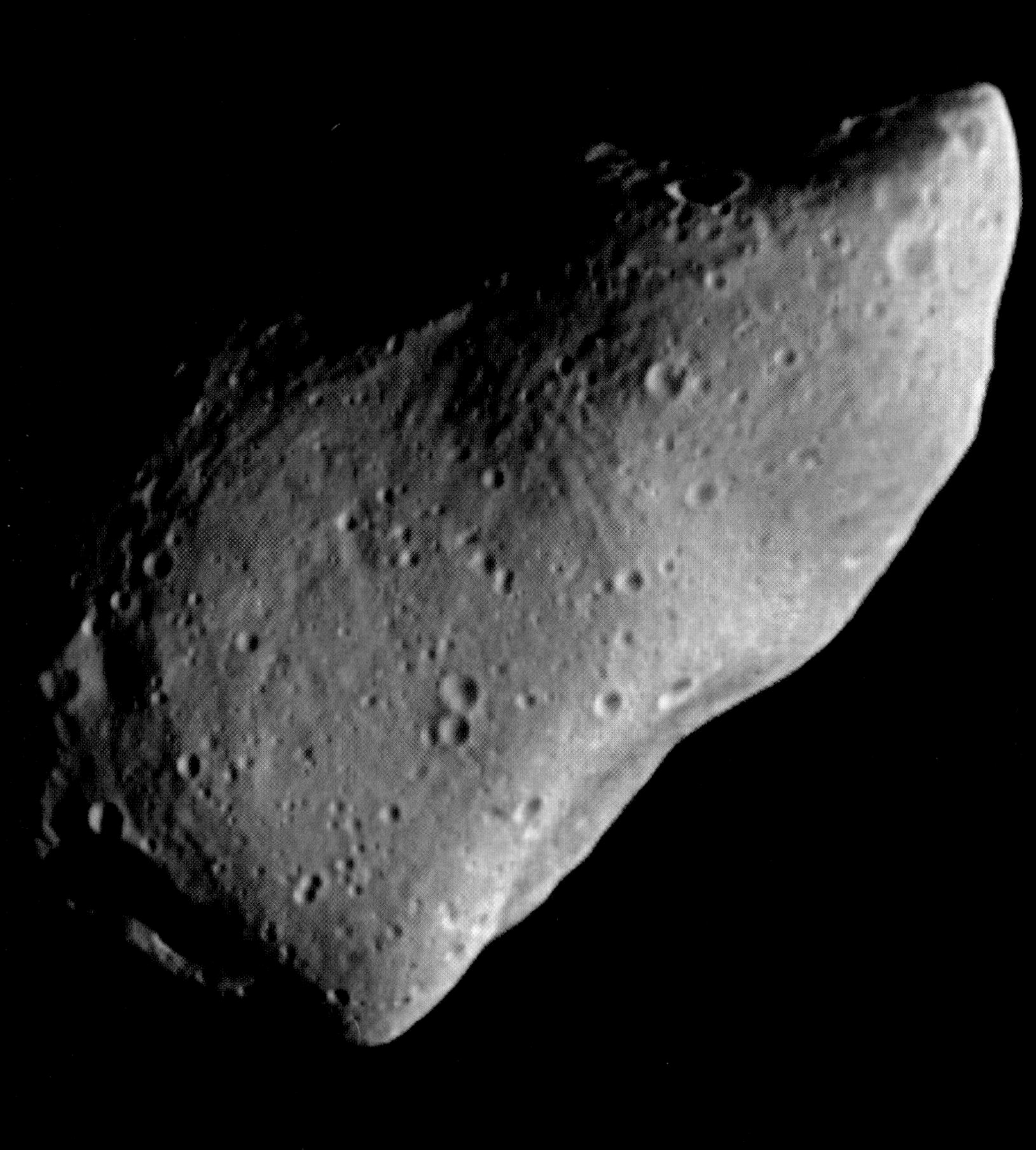

Asteroids are sometimes called minor planets because they can be hundreds of miles across.

Asteroids

Italian astronomer Giuseppe Piazzi discovered the first and largest asteroid. Astronomers are scientists who study objects in space. In 1881, Piazzi was looking at the sky to find a "lost" planet. At that time, people believed a planet orbited between Mars and Jupiter. Piazzi saw a light in the sky that no one had ever seen. Every night, he watched the light as it moved through the sky.

At first, Piazzi believed that he had found the lost planet. But he realized the object was too small to be a planet. Piazzi named the newly discovered kind of object an asteroid. Asteroid means starlike. Piazzi named the first asteroid Ceres after the Roman goddess of agriculture. Many objects in space are named after characters from Roman myths. A myth is a story told to explain something in nature.

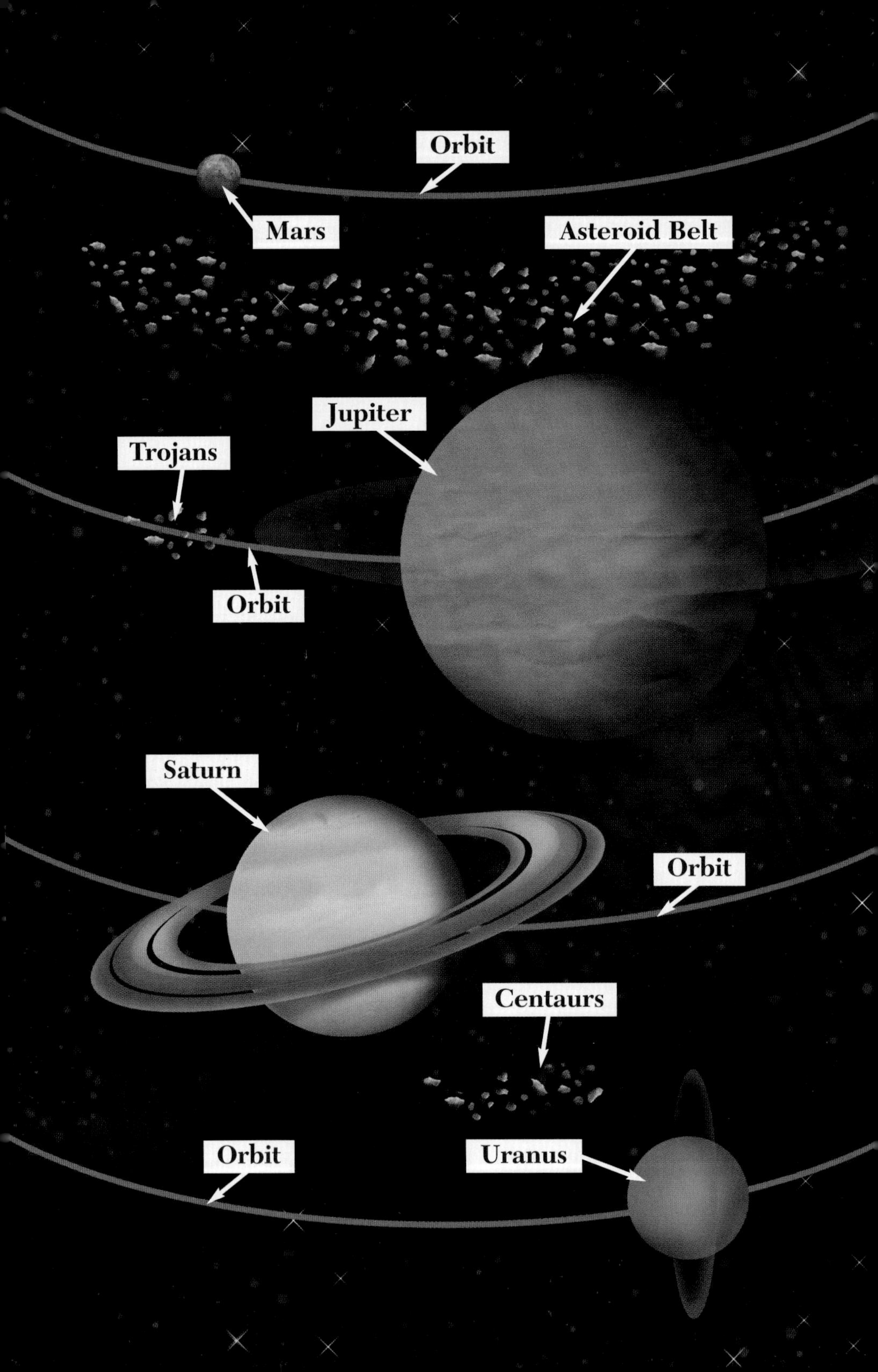

Orbit
Mars
Asteroid Belt
Jupiter
Trojans
Orbit
Saturn
Orbit
Centaurs
Orbit
Uranus

 **Most asteroids are in the main asteroid belt between Mars and Jupiter.**

Asteroid Belts

Asteroids are in many different places in the solar system. Some asteroids are close to the Sun. One group of asteroids is called the Apollos. This group orbits between Earth and Mars in the inner solar system.

The largest group of asteroids lies between Mars and Jupiter. Astronomers call the thousands of asteroids there the asteroid belt. Asteroids in this belt are between 170 million and 370 million miles (274 million and 595 million km) from the Sun.

Asteroids in the outer solar system are far away from the Sun. Another group of asteroids, called the Trojans, shares Jupiter's orbit around the Sun. Still farther from the Sun are the Centaurs, which travel between Saturn and Uranus.

Astronomers are not sure if the Centaurs are asteroids or comets. Comet tails show only when comets are close enough for the Sun to warm their gases. Beyond Saturn, comets stay very cold. At that distance, it is hard for astronomers to tell asteroids and comets apart.

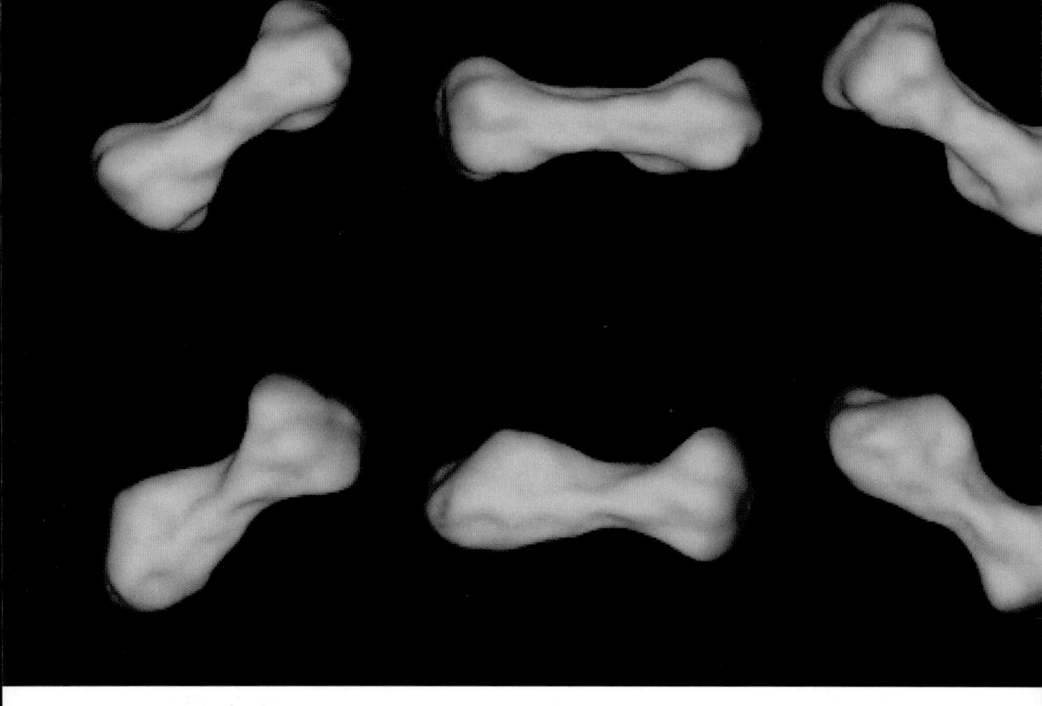

These are different views of Kleopatra, an asteroid that looks like a giant bone.

What Asteroids Look Like

Asteroids have a wide range of sizes. Ceres is the largest asteroid. It is almost 600 miles (966 km) wide. This is one-fourth the size of Earth's Moon. Smaller asteroids are less than one-half mile (0.8 km) wide.

Asteroids can be many different shapes. Some asteroids are round. Others are lumpy ovals. Ida is an asteroid that looks like a giant potato. The asteroid Kleopatra looks like a large dog bone.

Asteroids have rocky surfaces. Bumps, ridges, and holes called craters dot their surfaces. Small rocks called meteorites often crash into asteroids. The crashes cause explosions. These explosions leave craters when they blast some of the asteroids' rock into space. The small pieces of rock blasted into space become meteoroids.

Types of Asteroids

There are many kinds of asteroids. The three most common kinds of asteroids are S-type, M-type, and C-type. Each kind of asteroid is made up of different things.

S-type asteroids are made of rock. They look reddish gray because they are mostly made of stone. S-type asteroids orbit close to the Sun near the inside of the main asteroid belt.

Nickel and iron make up M-type asteroids. These asteroids look silver-gray. The metals in these asteroids reflect sunlight well. So they are the brightest kind of asteroid. M-type asteroids orbit in the middle of the main asteroid belt.

C-type asteroids are made mostly of carbon materials. They are darker than S-type asteroids because carbon does not reflect sunlight well. C-type asteroids orbit far from the Sun in the outer edges of the main asteroid belt. Most asteroids are C-types.

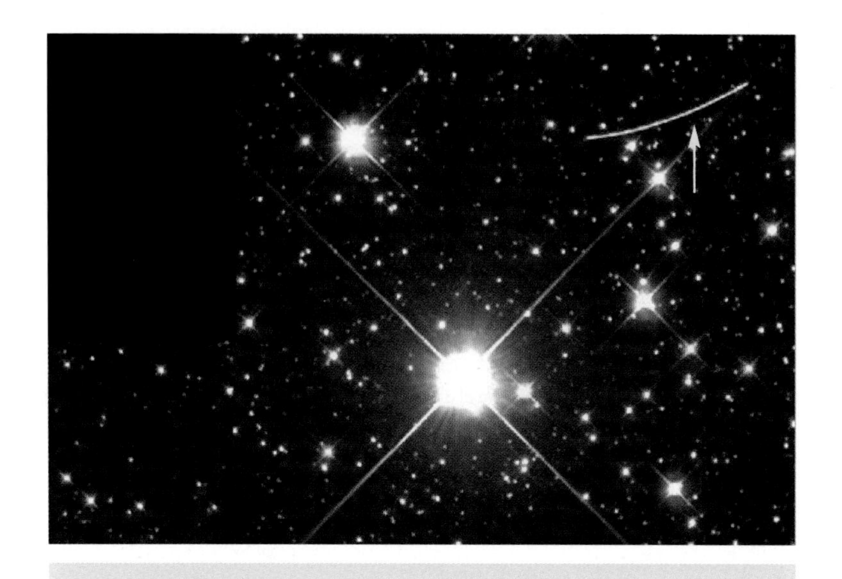

Asteroids reflect light from the Sun. Through telescopes, an asteroid's movement shows up as a streak of light called an asteroid trail. The telescope did not take a picture of the black area in this photograph.

Orbits and Gravity

Asteroids are usually very far apart from each other. Many asteroids travel through space alone for millions of years. But sometimes asteroids crash into each other. When that happens, they break apart. Pieces go in all directions. Large asteroids can break up into a number of small asteroids and meteoroids.

The gravitational force of an asteroid and a planet will pull on each other. Planets are larger, denser, and have a stronger gravitational force than asteroids.

Tiny Moon

Asteroids can be large enough to have moons. A spacecraft on its way to Jupiter passed the asteroid Ida. The asteroid is 32 miles (52 km) long. Pictures sent back by the spacecraft showed that Ida has its own moon. A tiny chunk of space rock is orbiting it. The rock may have exploded from Ida when another asteroid or meteor crashed into it.

Sometimes a planet's gravitational pull will make asteroids crash into the planet.

Other asteroids may be moving fast enough or be large enough to escape from a planet's gravitational pull. They may avoid crashing into the planet. But because the gravitational pull drew the asteroid out of its original orbit and closer to the planet, the shape of the asteroid's orbit changes.

The new orbit may bring the asteroid closer to the Sun. In time, these new orbits may also cause the asteroids to come near another planet. The gravitational pull from one of these planets may change the asteroid's orbit again. It may eventually crash into a planet. Or it may travel out of the solar system.

This is a photograph of a comet streaking through the sky over Mono Lake in California.

Comets

Comets are chunks of ice with frozen gases, dust, and rock mixed in. American astronomer Fred Whipple described comets very well. He said they are like dirty snowballs shooting through space.

Comets come from the outer solar system. Some comets come from the Kuiper Belt. This area of space just beyond Neptune is home to many comets. Most comets come from the Oort Cloud. This frozen cloud of gases and dust is trillions of miles away at the edge of the solar system. Some astronomers think that billions of comets travel in this cloud.

Sometimes the gravitational pull from a passing star changes a comet's orbit. A comet may then travel toward the Sun. People on Earth see the comet only when it enters the inner solar system and reflects light from the Sun.

This photograph shows a comet streaking across the starry sky.

Parts of a Comet

Comets are made up of several parts. A comet in cold, deep space is just an iceball. The iceball is made up of water, dust, rock, and frozen gases, such as ammonia and methane.

As a comet passes Jupiter, the Sun's heat begins to warm it. Some of the ice melts and releases gases that were frozen. The gases form a glowing shell around the iceball. This shell of gases is called a coma. Large comas can be 600,000 miles (1 million km) across. The icy center of a comet is its nucleus. The nucleus is much smaller than the coma. It may be only a few miles wide. Together, the nucleus and the coma are called the head of a comet.

The Sun's energy blows away from the Sun in the solar wind. The solar wind blows some of the gas and dust away from the head of the comet. This makes a tail. Tails may stretch millions of miles into space. Comets may even have more than one tail.

There are two types of comet tails. A type I tail is mostly gas. It streaks away from a comet in a straight line. Type I tails appear blue. Type II tails are mostly dust. These tails curve away from a comet. They appear yellow.

Comets cool down again as they move away from the Sun. The water and gas in the comet freeze. No more gas is released. The tail disappears, and the comet becomes just an iceball again. The comet stays in its iceball state until it travels back toward the Sun.

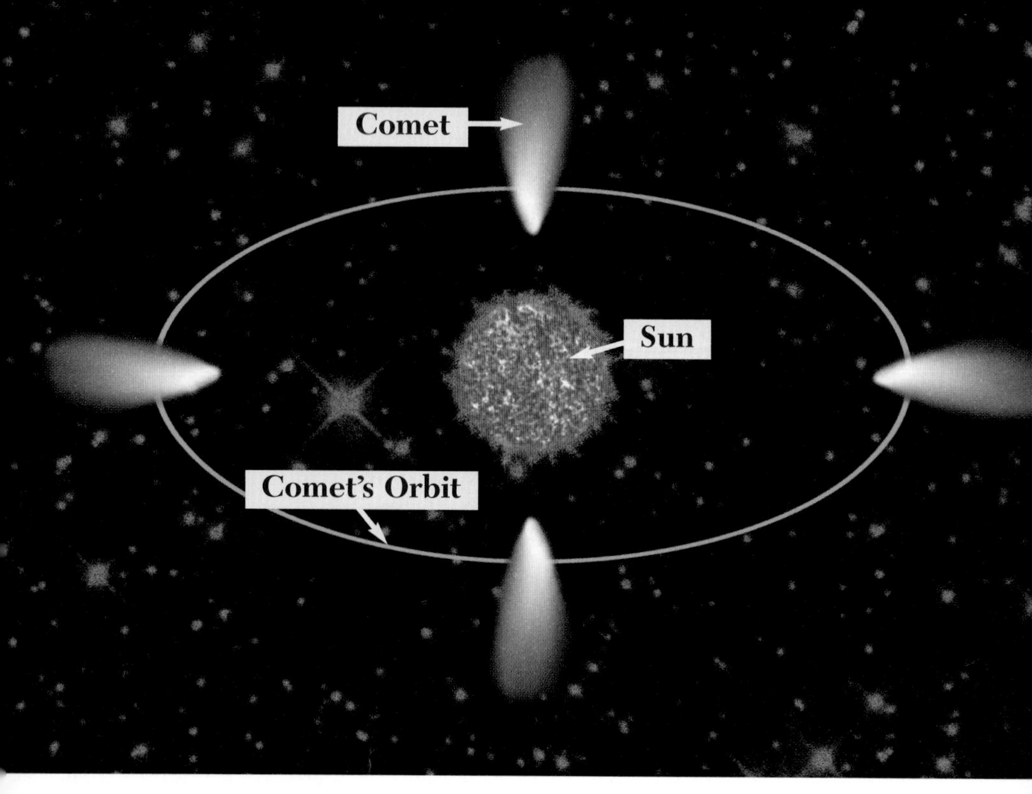

Comet Orbits

The shape of a comet's orbit is elliptical. The word elliptical means like a stretched-out circle. Some comets have huge orbits that cover trillions of miles of space. These comets are called long-period comets. They travel from the edge of the solar system to near its center close to the Sun. It may take long-period comets several thousand years to complete an orbit. Most long-period comets are from the Oort Cloud.

Sometimes a planet's gravitational pull traps a comet into a new, smaller orbit before it can return

Halley's Comet

Halley's comet is the most famous short-period comet. Since about 240 B.C., its orbit has brought it traveling back around the Sun every 76 years. Astronomers named the comet after English astronomer Edmond Halley. Halley studied the path of a comet he saw in 1682. He realized that comets seen in 1607 and 1531 were really the same object.

to deep space or the Oort Cloud. Then the comet travels between the planet and the Sun. These comets are called short-period comets. Short-period comets return to the Sun every 200 years or less. Comets in the Kuiper Belt are also short-period comets.

A comet's speed changes as it travels in its orbit. A comet heading toward the Sun begins to pick up speed as it is pulled by the Sun's gravity. As it gets close to the Sun, it moves very fast. The comet swings around the Sun and then heads back out into space. It slows down until it starts its return to the Sun.

The comet's tail also changes as it travels in its orbit. It always points away from the Sun. The tail grows as the comet nears the Sun. The Sun's energy pushes the tail behind the comet. Its tail is pushed in front when the comet moves away from the Sun.

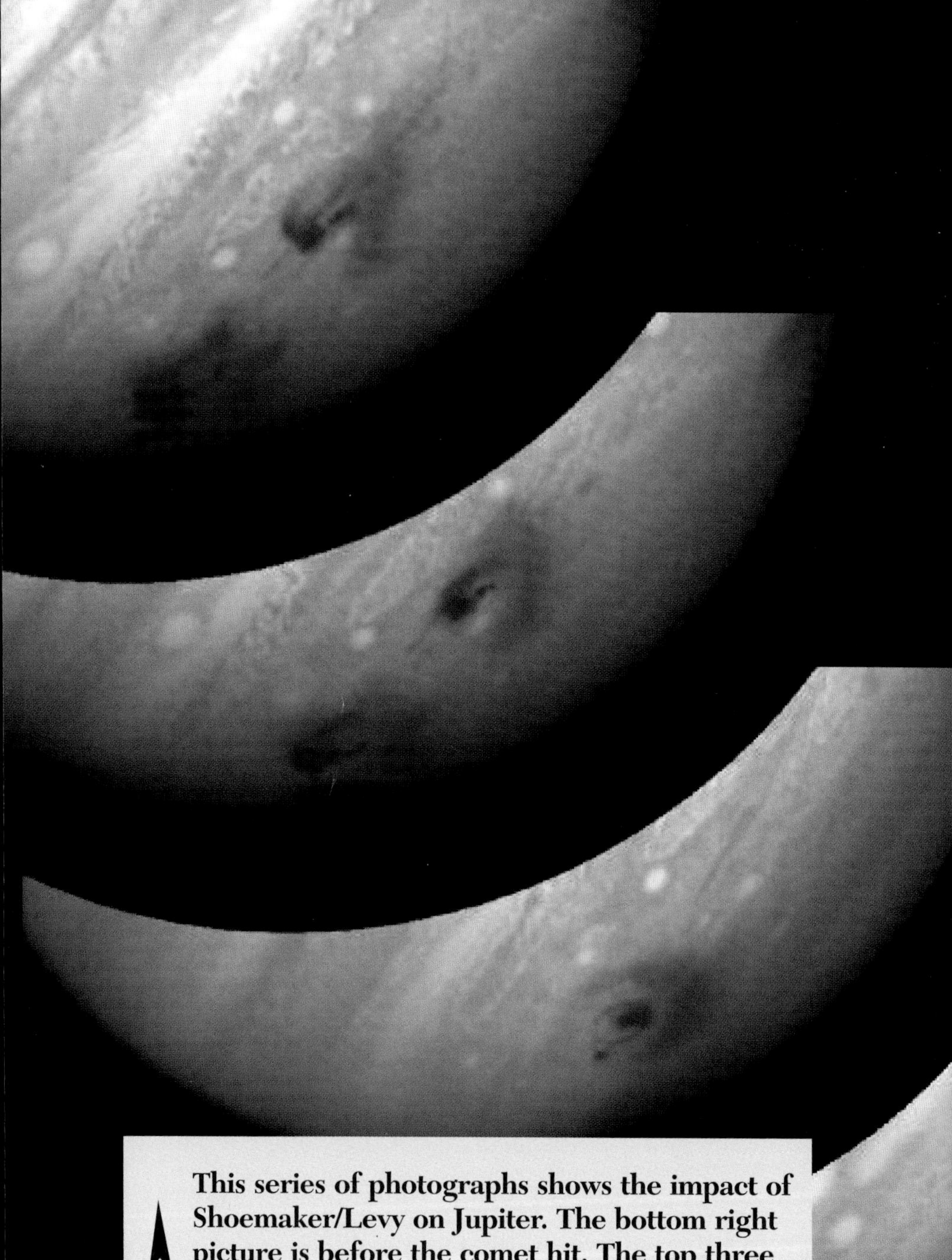

This series of photographs shows the impact of Shoemaker/Levy on Jupiter. The bottom right picture is before the comet hit. The top three pictures show the spots left by the explosions.

Comet Shoemaker-Levy

In 1993, astronomers Carolyn and Eugene Shoemaker and David Levy spotted a comet with an unusual appearance. Close-up pictures showed that the comet was actually 21 small comets traveling in a straight line. The small comets came from one large comet. The comet was named Shoemaker/Levy after the astronomers who discovered it. Jupiter's strong gravity pulled the large comet apart. Some of the larger pieces were 2.5 miles (4 km) across.

Shoemaker/Levy's orbit brought the pieces back toward Jupiter. In July 1994, the pieces of the comet entered Jupiter's atmosphere. An atmosphere is a layer of gases that surrounds a planet. The 21 pieces exploded into fireballs when they came into contact with Jupiter's gases.

Over the next several days, the fireballs smashed into Jupiter at speeds of 37 miles (59.5 km) per second. Each piece produced a huge explosion that sent streams of Jupiter's gas exploding into space. The comet crashes left dark spots in the planet. It took several months for the spots to disappear.

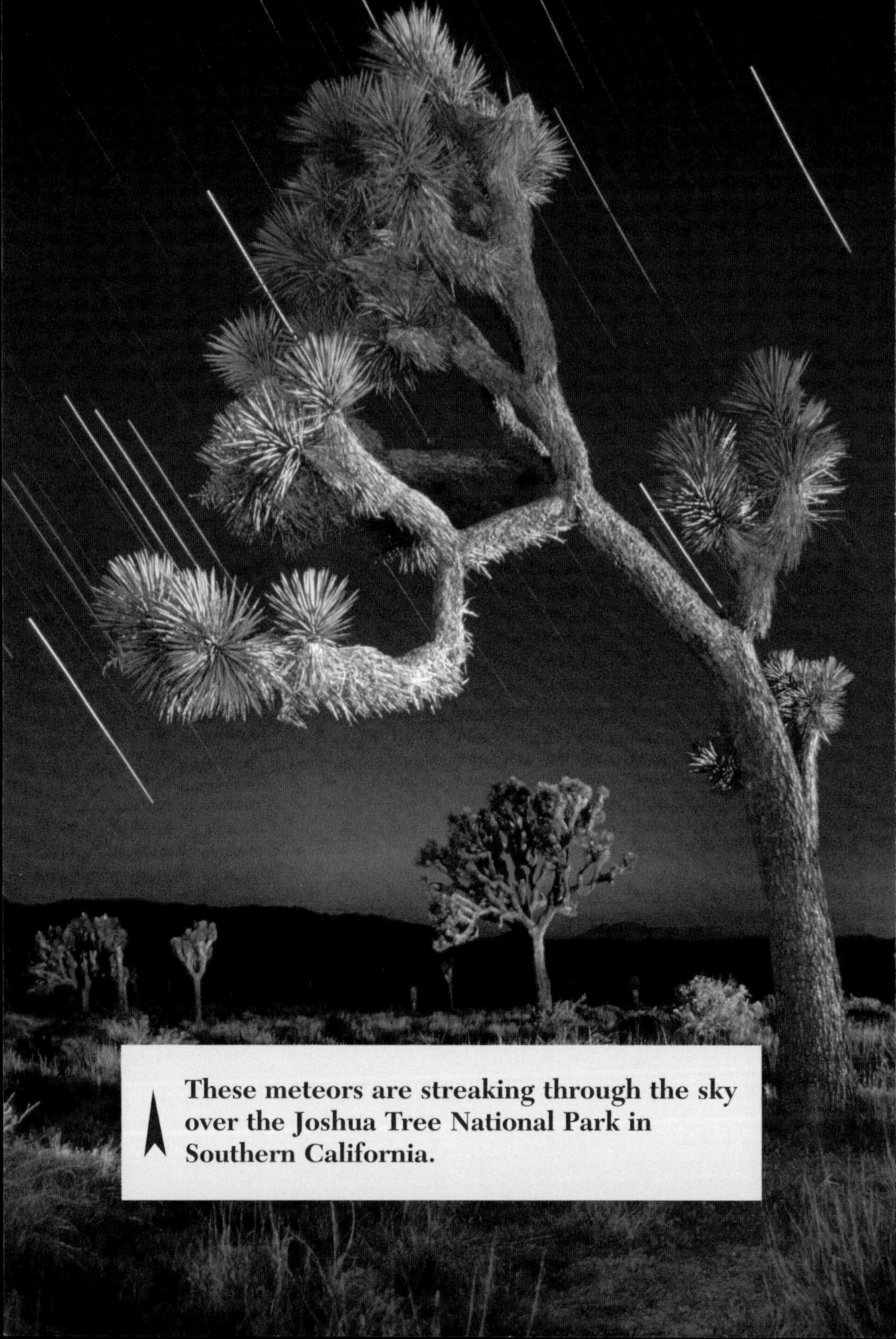

These meteors are streaking through the sky over the Joshua Tree National Park in Southern California.

Meteoroids

Meteoroids are dust grains or small pieces of metal and rocks in space. They may be dust from comets or chips from asteroids, moons, or other planets. Like everything else in the solar system, meteoroids orbit the Sun.

Some meteoroids enter Earth's atmosphere. Meteoroids are called meteors as soon as they enter Earth's atmosphere. Earth's atmosphere completely destroys most meteors. They break into small pieces and burn. Then they look like streaks of light in the sky. They are often called shooting stars.

Some meteors survive Earth's atmosphere and fall to Earth's surface. Meteors are called meteorites as soon as they crash into a planet or other object in space.

Types of Meteoroids

There are several types of meteoroids. Each type is made up of different materials. Stony meteoroids are more than 80% rock. Most meteoroids are stony. Iron meteoroids are made mostly of metals, such as iron and nickel. They are usually 90% iron, 9% nickel, and 1% other metals.

Stony iron meteoroids are the rarest type of meteoroid. They are made up half of iron and half of rock. Only about 1% of meteoroids are stony iron.

Stony and iron meteoroids come from asteroids. When asteroids crash together, pieces break off. The pieces fly in many directions and become meteoroids.

Sometimes meteorites crash into planets or moons. This causes explosions that send some rocks shooting into space. These rocks become meteoroids. After time, they may crash into other objects in space. They are called meteorites as soon as they crash into another object.

This stony meteorite came from Mars. Scientists took this picture with a ruler to show how big the rock is.

A person is holding a slice of a stony iron meteorite in front of a flashlight to better show the crystals inside the meteorite.

What Meteoroids Look Like

Meteoroids can be many different sizes. Some meteoroids are as large as houses, and some are as small as baseballs. Most meteoroids are tiny. They are no bigger than a grain of sand or speck of dust.

Meteoroids are different from rocks on Earth. All meteoroids contain some metal. They are often attracted to a magnet. Most Earth rocks have no metal in them, and magnets do not attract them. Most meteoroids contain nickel, but Earth rocks contain very little nickel. Meteoroids are also denser than Earth rocks.

Meteoroids that have fallen to Earth have different surfaces. Some meteorites are smooth. Others have many little dents in them. Meteorites that have just fallen to Earth have black crusty surfaces. The black crust is the melted outer surface. The surface burned as it fell through Earth's atmosphere.

Meteoroids contain many different minerals and metals. The inside of each meteoroid looks different depending on what it is made of. Some stony iron meteorites have crystals inside them. Many stony meteoroids contain olivine. This hard, greenish rock forms under high heat and pressure. Some stony meteoroids have diamonds in them.

Fireballs are caused when large meteoroids enter Earth's atmosphere.

Shooting Stars

Scientists believe millions of meteoroids enter Earth's atmosphere each year. These meteors travel up to 20 miles (32 km) per second. Passing through an atmosphere causes friction. Friction is heat caused by two things rubbing against each other. In a second or two, the heat begins to destroy the meteor.

A burning meteor looks like a bright flash. It streaks across the sky for only a second or two. People have nicknamed these flashes of light shooting stars.

The color of the shooting star depends on the temperature of the meteor and what it is made of. The flashes may look white, yellow, blue, green, orange, or red.

The brightest meteors are called fireballs. People can see fireballs for hundreds of miles. Sometimes fireballs explode. These exploding fireballs are called bolides. Bolides are like bombs. The explosions cause loud noises.

The number of meteors in the sky changes. Usually about 4 to 10 meteors streak through the sky each hour. But sometimes Earth will pass through the path of a comet. Hundreds of leftover dust specks from the comet enter Earth's atmosphere. Each dust speck causes a bright streak in the sky. This causes a meteor shower.

During a meteor shower, several meteors may flash every minute. At other times, thousands of meteors per hour streak through the sky. These are called meteor storms. Meteor storms rarely happen.

Meteorite Craters

Meteors that survive the trip through Earth's atmosphere fall to Earth as meteorites. Meteorites make craters when they crash into objects in space. The planets Mercury, Venus, Earth, and Mars have craters. Most moons in the solar system have craters, too. Earth's moon is covered with craters.

Small meteorites are the size of pebbles. They make cup-sized round craters when they smash into the ground. The force from the crash pushes up the edges of the hole. The meteorite rests in the middle of the crater.

A large meteorite makes a deep crater that can be hundreds of miles across. It blasts dirt and rocks into the air when it crashes. The meteorite breaks apart when it explodes. Heat from the explosion will sometimes melt the surface rock. The melted rock may splash up in the center of the crater. When it hardens again, it forms a small mountain in the center of the crater.

Astronomers can tell the difference between old and new craters. Wind and rain have worn down old craters. Sometimes new meteorites strike old craters. This breaks the old crater walls and makes fresh craters on top of the old ones.

Billions of years ago, a large meteorite struck the Moon. It made this crater, which is about 60 miles (97 km) wide.

This overhead picture shows what Meteor Crater in Arizona looks like.

Meteor Crater, Arizona

Arizona is the best place to see a meteorite crater up close. About 50,000 years ago, a meteorite struck the desert near Winslow, Arizona. The meteorite weighed about 1 million tons (1,016,047 metric tons).

The meteorite crash made a crater in the desert that is 4,000 feet (1,219 m) wide and 600 feet (183 m) deep. The explosion destroyed most of the meteorite. But some pieces were left over. The pieces were made out of iron. The meteorite survived its trip through the atmosphere because it was made of metal.

Scientists have carefully studied the crater. They discovered a bent crater rim. The rim stands about 200 feet (60 m) above the flat desert. The force of the blast tipped the rock rim upward. Scientists also found deep underground cracks spreading out from the crater. Scientists think an object this size or greater crashes into Earth once about every 100 years.

Tunguska, Siberia

On June 30, 1908, people in Tunguska, Siberia, saw a white streak in the sky. Suddenly, a huge explosion lit up the sky. The explosion might have been a bolide. Or it could have been a comet that exploded in the air just above the ground.

The explosion knocked people 60 miles (97 km) away off their feet. It flattened all the trees in a 20-mile (32-km) area. People heard the noise from the blast from hundreds of miles away. Luckily, no one lived where the comet fell, so no one died.

Earth and Moon Craters

Earth has only a few dozen meteorite craters because its surface is about 75% water and only 25% land. Most meteorites hitting Earth fall into the oceans. This makes a large splash but no crater. The water slows the meteorite down so the explosion is not as strong.

Scientists think Earth has had more craters in the past. Water and wind wear down craters on Earth. This process is called erosion. Water fills in some craters to make lakes. Wind blows dirt and dust into some craters until they are filled in. Over time, the craters disappear.

Erosion has worn down the Manicougan Crater in Quebec, Canada. A large meteorite hit the surface there about 212 million years ago. The explosion made a 60-mile (97-km) wide crater. Over millions of years, water and wind have filled in most of the crater. Today, the crater is a ring-shaped lake.

The Moon has millions of craters. It has so many craters because it is very different from Earth. The Moon has no atmosphere or liquid water on its surface. There is no wind, rain, or water on the Moon, so there is no erosion. Craters there stay the same for millions of years.

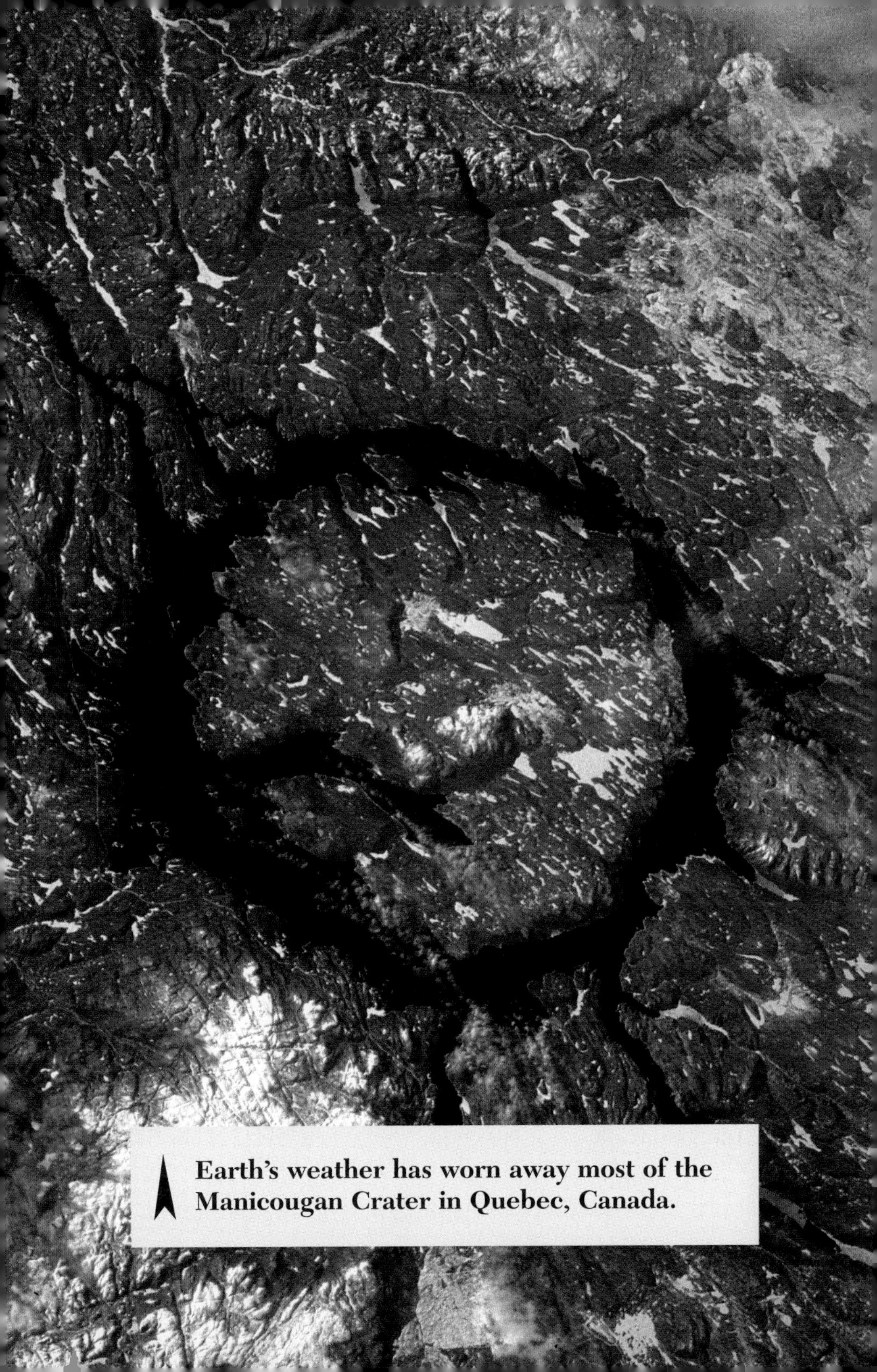

Earth's weather has worn away most of the Manicougan Crater in Quebec, Canada.

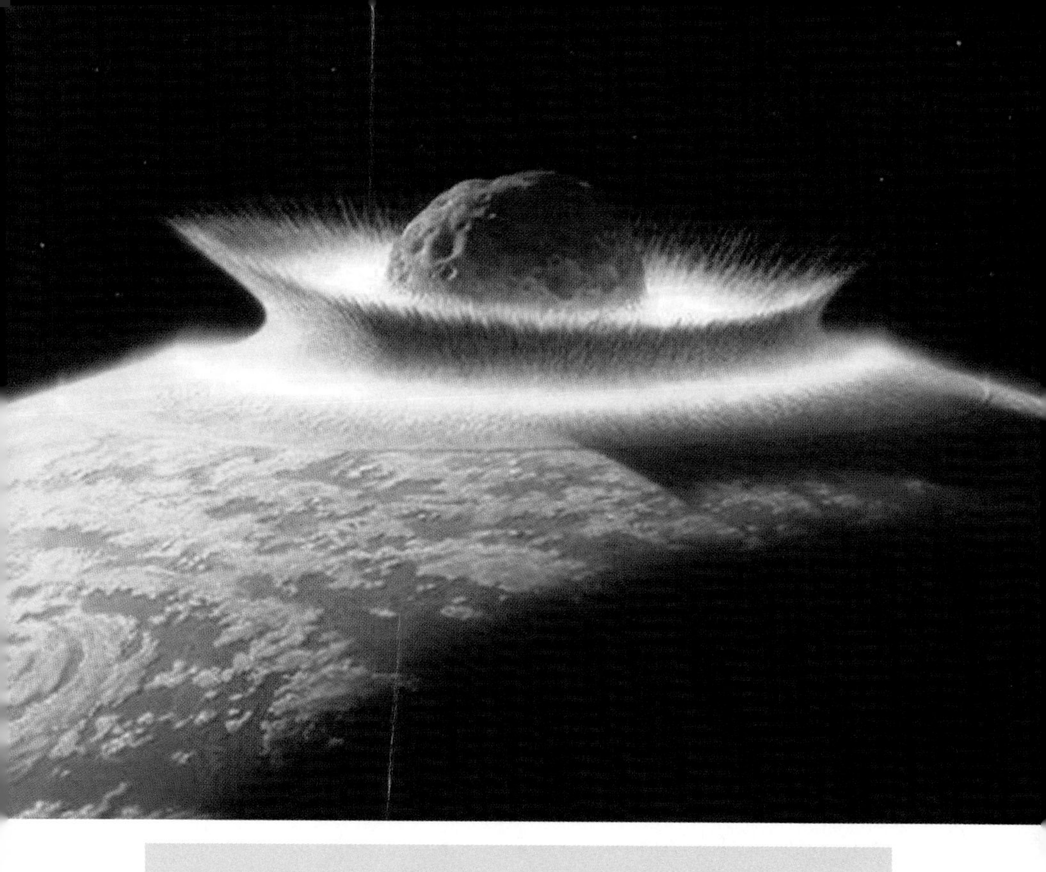

Many experts think a large asteroid or comet struck Earth and killed the dinosaurs. This is an artist's idea of what the crash looked like.

End of the Dinosaurs

Dinosaurs died out 65 million years ago. For many years, scientists have tried to find out what killed the dinosaurs.

Many scientists now think an asteroid struck Earth when the dinosaurs lived. The explosion released gases and dust. The dust rose in the atmosphere and blocked sunlight. The sky darkened.

The air became very cold. Gases from the explosion mixed with rain to make it poisonous. Many plants and animals died. The world warmed again when the dust settled. But by then all the dinosaurs were dead.

Scientists believe the crater left by the asteroid is in Mexico. Half of the crater lies in the Gulf of Mexico and half on the Yucatan peninsula. Circular cracks stretch out from the crater. The crater has eroded over the years. But it is still more than 112 miles (180 km) wide.

The asteroid that struck Mexico was 6 miles (10 km) wide. Scientists think an object this size crashes into Earth about every 300,000 years.

Lunar and Martian Meteorites

Some scientists look for meteorites. They study meteorites to find out what the meteorites are made of and where the rocks came from.

Scientists have found meteorites from Mars and the Moon. One special meteorite found in Antarctica came from Mars. The meteorite has tubelike objects inside it. Some scientists think these are fossils of ancient Martian life. A fossil is the remains of animal or plant life. Other scientists do not think that the the meteorite contains signs of Martian life.

Scientists sometimes use robots to pick up meteorites in Antarctica.

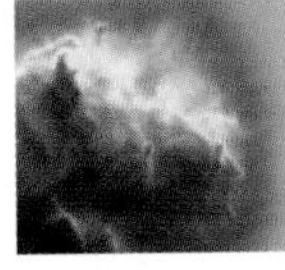

The Search for Asteroids and Comets

Scientists believe that a comet or another huge asteroid might hit Earth some day. Explosions made when these objects crash could cause great damage. They would kill many people.

The U.S. government and other educational agencies fund a space program called Spacewatch. Scientists working for Spacewatch look for asteroids and comets that might be heading for Earth. Scientists measure the speed of every comet or asteroid they find. They find out what direction the objects are traveling in. A computer uses this information to make a map of each object's orbit for the next several years.

***NEAR Shoemaker* took this picture of the asteroid Eros.**

Saving the Earth

Astronomers have a list of the orbits of known asteroids and comets. They are tracking these objects' orbits. They add any new asteroids or comets they find to their list.

So far, astronomers know of no asteroid or comet heading toward Earth. But this might change. Scientists have thought of ways to try and protect

Earth. People could send rockets with bombs toward the objects. The bombs would explode on or near the objects. The explosions could push the large objects into different orbits away from Earth. Or the explosions would break the large objects into small pieces. The small pieces would still hit Earth. But they would not cause as much damage as one large object.

Scientists are also coming up with ways to use asteroids and comets. People on Earth may someday need rock and metal from asteroids to make buildings and space stations. They could get water from comets and turn it into fuel and oxygen.

Someday, special spaceships will be sent to nearby asteroids and comets. Mining machines will remove water, metals, and rock. Spaceships will take these materials where they are needed.

Spacecraft help gather new information about asteroids and comets. *Giotto,* launched in 1985, took pictures of Comet Halley's nucleus. *NEAR Shoemaker* was launched in 1997. It is orbiting the nearby asteroid, Eros.

Scientists hope to find out more about how the solar system formed by studying asteroids, comets, and meteoroids. Maybe they will discover more about Earth, its history, and its future.

Glossary

asteroid (AS-tuh-roid)—giant space rock

astronomer (ah-STRAHN-uh-mur)—a scientist who studies objects in space

atmosphere (AT-muh-sfear)—a layer of gases that surrounds an object in space

bolide (BOH-lide)—an exploding fireball

coma (COH-mah)—the glowing shell of gases that sometimes surrounds a comet's nucleus

comet (KOM-it)—a large ball of ice, dust, and rock that has a tail of glowing gases when it passes near the Sun

meteor (MEE-tee-ur)—a meteoroid that has entered a planet's atmosphere; meteors look like flashes of light streaking across the sky.

meteorite (MEE-tee-ur-rite)—a meteoroid that crashes into an object in space

meteoroid (MEE-tee-ur-OYD)—a piece of rock, rock and metal, or dust in space

nebula (NEB-yoo-lah)—a huge cloud of gas and dust in space

nucleus (NOO-klee-uhs)—the solid, icy center part of a comet

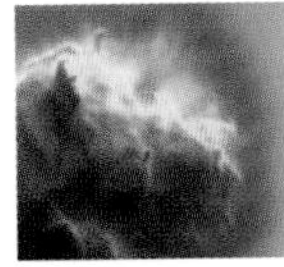

Internet Sites and Addresses

Comets
http://www.windows.umich.edu/comets/comets.html

NASA for Kids
http://kids.msfc.nasa.gov

North American Meteor Network
http://web.infoave.net/~meteorobs

Star Child: A Learning Center for Young Astronomers
http://starchild.gsfc.nasa.gov/docs/StarChild/StarChild.html

Astronomical Society of the Pacific
390 Ashton Avenue
San Francisco, CA 94112

NASA Headquarters
Washington, DC 20546-0001

Index